α Alpha

Focus: Single-Digit Addition and Subtraction

Test and Activity Booklet

1·888·854·MATH (6284) - MathUSee.com
Sales@MathUSee.com

# Math·U·See®

Sales@MathUSee.com · 1·888·854·MATH (6284) · MathUSee.com
Copyright © 2010 by Steven P. Demme

Graphic Design by Christine Minnich. Illustrations by Gregory Snader.

All rights reserved. No part of this book may be reproduced, stored
in a retrieval system, or transmitted in any form by any means—electronic,
mechanical, photocopying, recording, or otherwise.

In other words, thou shalt not steal.

Printed in the United States of America

EXTRA FUN ACTIVITIES

This booklet contains extra activity pages for the student as well as the tests. See the next page for information about the activity pages. Go to page 73 to find the *Alpha* tests.

EXTRA ACTIVITY PAGES

In the first section of this booklet, you will find an extra activity page for each lesson. We hope that these pages provide an enjoyable way for students to practice what they have been learning in the lessons. (There are no solution pages for the extra activities.)

The concepts taught in *Alpha* are important to a student's future success in math. The focus of this level is memorizing addition and subtraction facts. We hope the variety of games and activities will enhance your student's experience with math and make it more enjoyable.

You may schedule the extra activity pages in any way that is useful to you. Some students may need a little help getting started with some of the activities. Be sure to keep it lighthearted and fun.

Check your instruction manual for more games and teaching tips.

Dot-to-dot activities are a great way to practice counting. Start at 1 and connect the dots to find the picture.

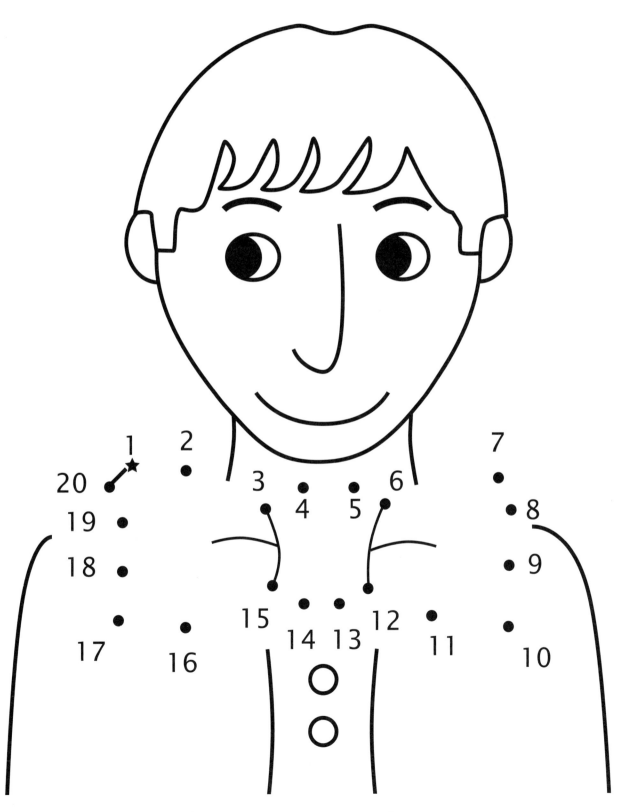

EXTRA FUN 2X

Start at 1 and connect the dots to find the picture.

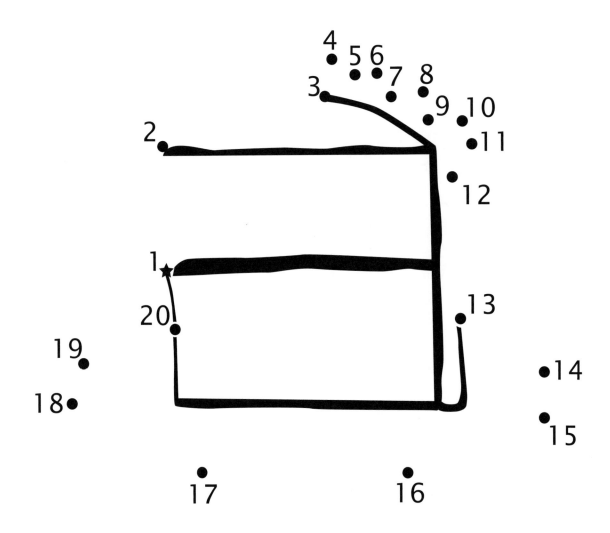

3X

Turn the paper sideways, and then place the bars over the corresponding pictures. Write the numbers in the blanks, and say each one. Color the pictures to match the blocks.

ALPHA EXTRA FUN

EXTRA FUN 3X

Match the block colors and color the house.
Try building a different house with your blocks.

Sammy had 0 toy trucks.
Draw a picture of Sammy's trucks.

Sammy's dad brought home 4 toy trucks.
Draw a picture of the trucks Dad brought home.

Write the number that tells how many toy trucks Sammy has now.

_____

EXTRA FUN 4X

Answer the silly word problems.
Can you find a block that matches each answer?

1. How many real elephants are in your house?

   _____

   If three elephants came to visit, how many elephants would there be in your house?

   _____

2. Tom saw zero giraffes in the front yard and zero giraffes in the back yard. How many giraffes did he see altogether?

   _____

3. How many ears do you have?

   _____

   How many more ears will grow on your head?

   _____

   How many ears will you have altogether?

   _____

4. Write the answer to the problem. Make up your own silly word problem and tell it to your teacher.

   6 + 0 = _____

Anna had five bananas. Mom gave her one more banana. Draw a picture that shows how many bananas Anna has now.

5 + 1 = _____

Sally has three cookies in her left hand and one cookie in her right hand. Draw a picture that shows how many cookies she has in all.

3 + 1 = _____

EXTRA FUN 5X

Find the block that shows one more. Write its name in the blank. You may color the blocks if you wish.

☐☐ 2 + 1 = _____

☐☐☐ 3 + 1 = _____

☐☐☐☐ 4 + 1 = _____

☐☐☐☐☐ 5 + 1 = _____

☐☐☐☐☐☐ 6 + 1 = _____

☐☐☐☐☐☐☐ 7 + 1 = _____

☐☐☐☐☐☐☐☐ 8 + 1 = _____

☐☐☐☐☐☐☐☐☐ 9 + 1 = _____

Skip count by 10. Start at the star and follow the dots to find a picture.

ALPHA   EXTRA FUN

19

To the parent: This 100 chart may be used as an aid while teaching counting and skip counting. It begins with zero as suggested by Math-U-See. It may be laminated for long-term use.

If you wish, have the student color the columns to match the block colors for the numbers in the unit places. For example, the first column would not be colored, the second column (1, 11, 21, 31, etc.) would be colored green, and so on.

0	1	2	3	4	5	6	7	8	9
10	11	12	13	14	15	16	17	18	19
20	21	22	23	24	25	26	27	28	29
30	31	32	33	34	35	36	37	38	39
40	41	42	43	44	45	46	47	48	49
50	51	52	53	54	55	56	57	58	59
60	61	62	63	64	65	66	67	68	69
70	71	72	73	74	75	76	77	78	79
80	81	82	83	84	85	86	87	88	89
90	91	92	93	94	95	96	97	98	99
100									

If the answer is 3, color the space brown.
If the answer is 4, color the space green.
If the answer is 5, color the space red.

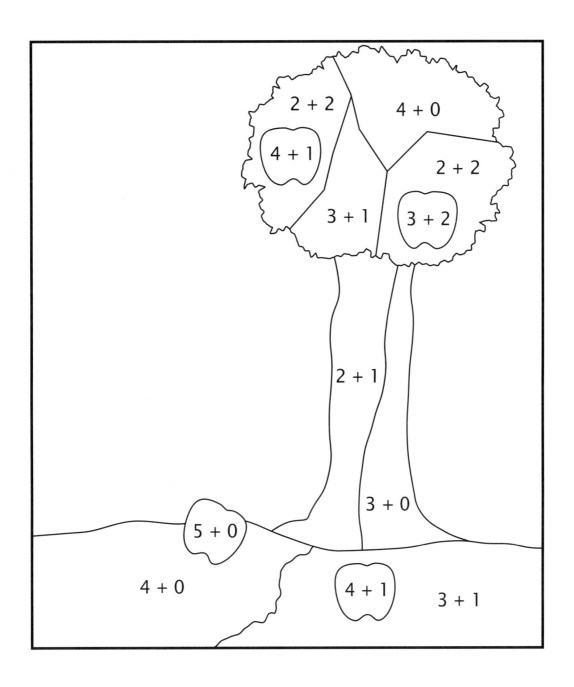

EXTRA FUN 7X

Fill in the names to make word problems about people you know. Write your answers in the boxes.

1. Last week, _____ earned five dollars doing chores.

   This week he didn't earn any money.

   How much did he earn in all?

   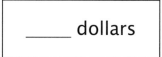
   _____ dollars

2. _____ is seven years old.

   How old will he or she be in two more years?

   _____ years

3. _____ built one block castle.

   _____ built three block castles.

   How many block castles have been built?

   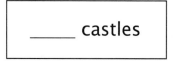
   _____ castles

4. Write the answer to the problem.

   $6 + 2 =$ _____

   Make up your own word problem and tell it to your teacher.

Draw pictures to help you solve a word problem with more than one step.

Jill made four pies.
Draw Jill's pies in the box.

Grace made two pies.
Draw Grace's pies in the box.

How many pies did the girls make in all? _____

The girls need eight pies.
Draw the pies you need to make eight pies in all.

How many more pies did you draw? _____

The dog ate three pies.
Cross out the pies the dog ate.

How many pies are left? _____

EXTRA FUN 8X

Skip count by 10. Start at the star and follow the dots to find the picture.

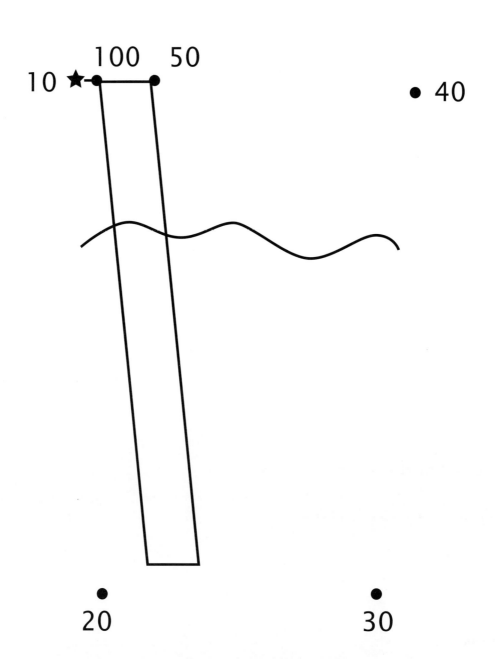

Start at 100 and connect the dots by counting *backwards* by ten.

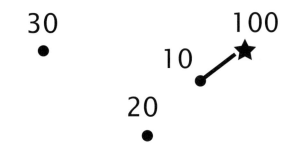

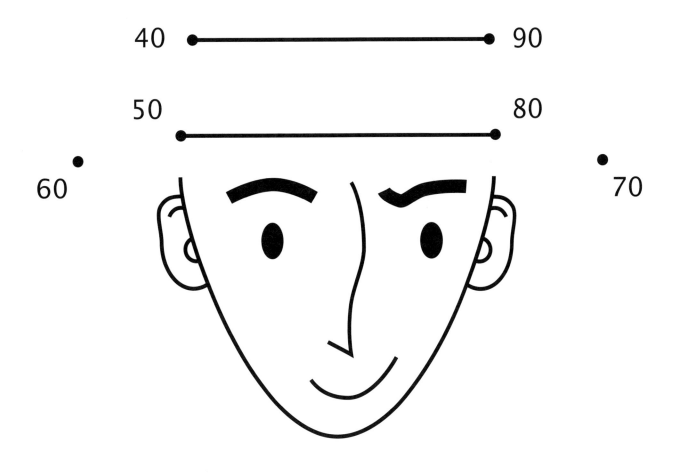

ALPHA EXTRA FUN

EXTRA FUN 9X

If the answer is 12, color the space black.
If the answer is 13, color the space yellow.
If the answer is 14, color the space red.
If the answer is 15, color the space green.
If the answer is 16, leave the space white.

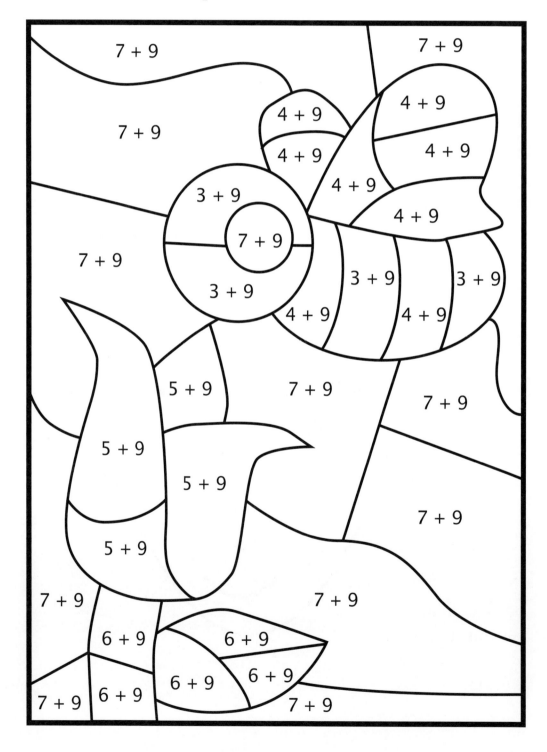

10X

Follow the directions.

Dan dropped 10 unit blocks on the kitchen floor.

The dog ate two blocks.
Cross out the blocks the dog ate.

There are _____ blocks left.

The baby hid two blocks.
Cross out the blocks the baby hid.

There are _____ blocks left.

Dan picked up two blocks and put them on the table.
Cross out the blocks Dan put on the table.

There are _____ blocks left.

Dan picked up two more blocks.
Cross out the blocks Dan picked up.

There are _____ blocks left.

Mom vacuumed two blocks.
Cross out the blocks Mom vacuumed.

There are _____ blocks left.

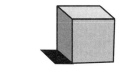
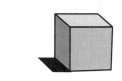
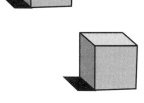
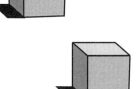

ALPHA EXTRA FUN

Follow the directions. This time we will start with nine instead of ten and count back by two.

Ava made nine cookies.

Dad ate two cookies.
Cross out the cookies Dad ate.

There are _____ cookies left.

Mom ate two cookies.
Cross out the cookies Mom ate.

There are _____ cookies left.

David ate two cookies.
Cross out the cookies David ate.

There are _____ cookies left.

The hamster ate two cookies.
Cross out the cookies the hamster ate.

There is _____ cookie left for Ava to eat.

11X

Skip count by two and connect the dots. After you have finished, try starting at 20 and counting backwards by two.

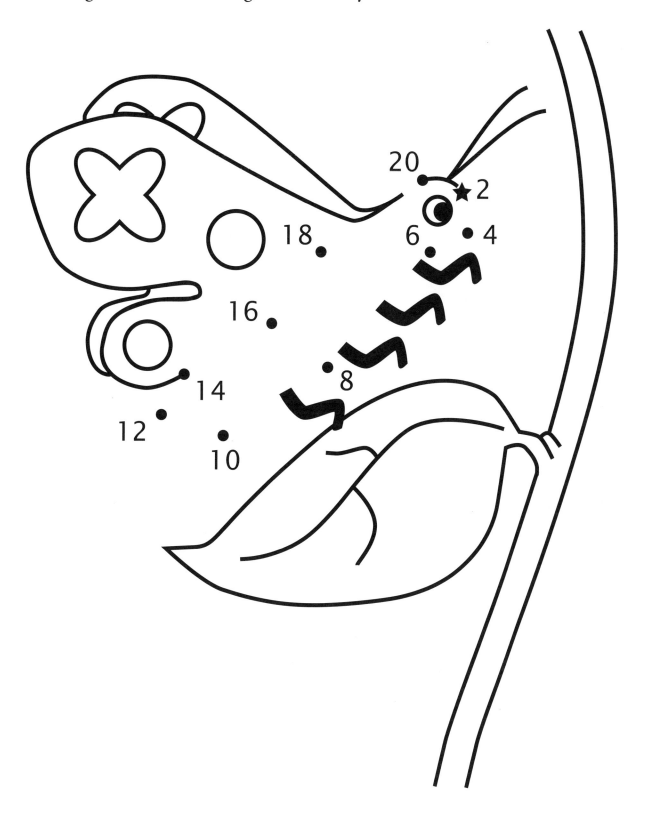

ALPHA   EXTRA FUN

EXTRA FUN 11X

Draw pictures to help you solve a word problem that has more than one step.

Riley blew up four red balloons.
Draw the red balloons in the box.

Riley blew up five green balloons.
Draw the green balloons in the box.

How many balloons does Riley have now? _____

Riley wants 12 balloons altogether.
Draw the balloons you need
to make 12 in all.

How many more balloons did you draw? _____

The cat broke eight balloons.
Cross out the balloons the cat broke.

How many balloons are left? _____

12X

Find the sums and use number words to write the answers in the boxes. The first one is done for you.

Across

1. 3 + 3 = 6

2. 4 + 5 = ___

3. 6 + 2 = ___

4. 3 + 7 = ___

5. 1 + 1 = ___

7. 2 + 3 = ___

8. 0 + 0 = ___

Down

1. 5 + 2 = ___

4. 1 + 2 = ___

6. 0 + 1 = ___

7. 2 + 2 = ___

ALPHA EXTRA FUN

EXTRA FUN 12X

If the answer is 8, color the space yellow.
If the answer is 10, color the space orange.
If the answer is 12, color the space red.
If the answer is 14, color the space brown.

Skip count by 5 to 50.

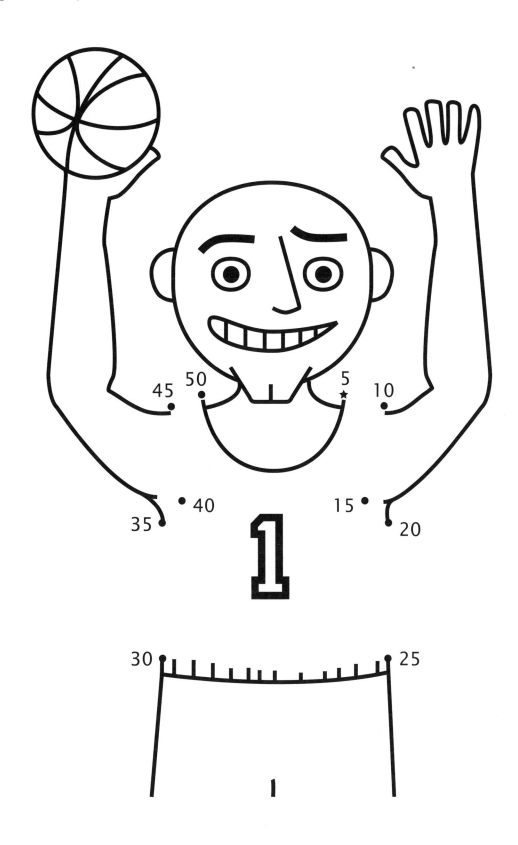

EXTRA FUN 13X

Start at 0 and add 2. Trace the line to the answer. Keep adding 2 and tracing the lines.

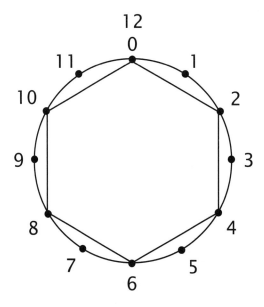

The shape inside the circle has six sides. It is called a hexagon.

Start at 0 and add 3. Draw a line to the answer. Keep adding 3 and drawing the lines.

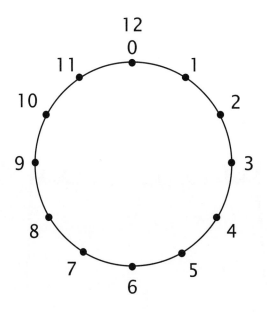

What shape did you draw inside the circle? _____

14X

Help us write the word problems. Write your answers in the boxes.

Fill in the blanks with kinds of fruit.

1. Julia had three _____ and three _____ .

    She found one _____ .

    How many pieces of fruit does she have now? ☐

Fill in the blanks with names of shapes.

2. Timothy drew four _____ and four _____ .

    Then he drew one _____ .

    How many shapes did he draw? ☐

Fill in the blanks with your favorite toys.

3. Peter bought five _____ and five _____ .

    Then he bought one _____ .

    How many toys did he buy? ☐

EXTRA FUN 14X

Start at 0 and add 4. Draw a line to the answer. Keep adding 4 and drawing the lines.

What shape did you draw inside the circle? _____

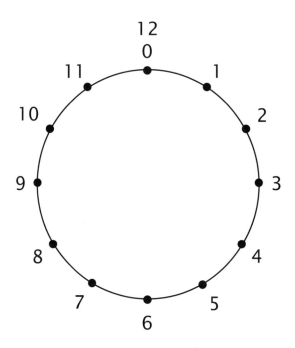

Challenge!

Start at 2 and add 4. Draw a line to the answer. Keep adding 4 to your answer and drawing the lines. The last time you add, count the dots to find the last answer.

What shape do you see now? _____

Find the missing numbers to practice making 10. The answers are on the back of each triangle.

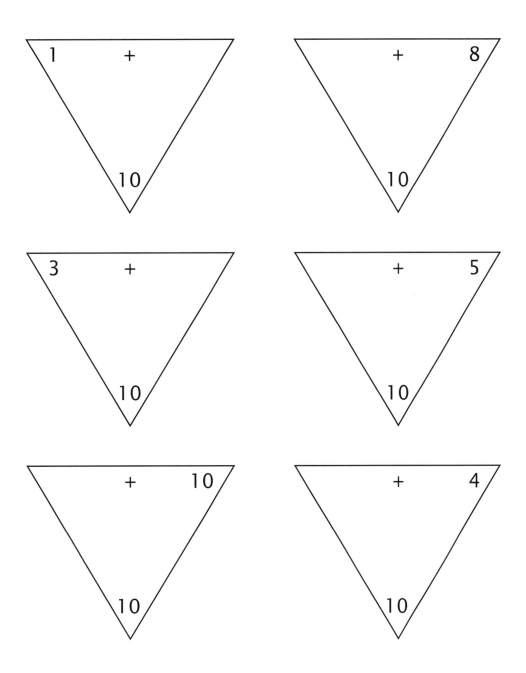

The triangles may be cut out and laminated if you wish. See the teaching tip in lesson 12 for more on using trios to teach addition.

EXTRA FUN 15X

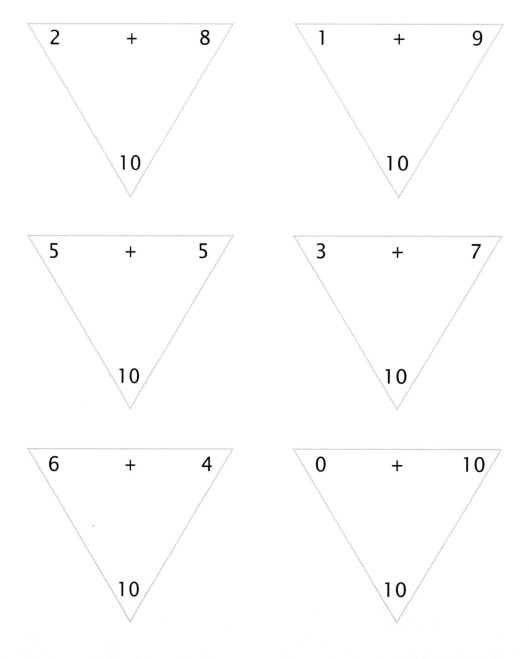

16X

These triangles will help you practice making nine. The answers are on the back of each triangle.

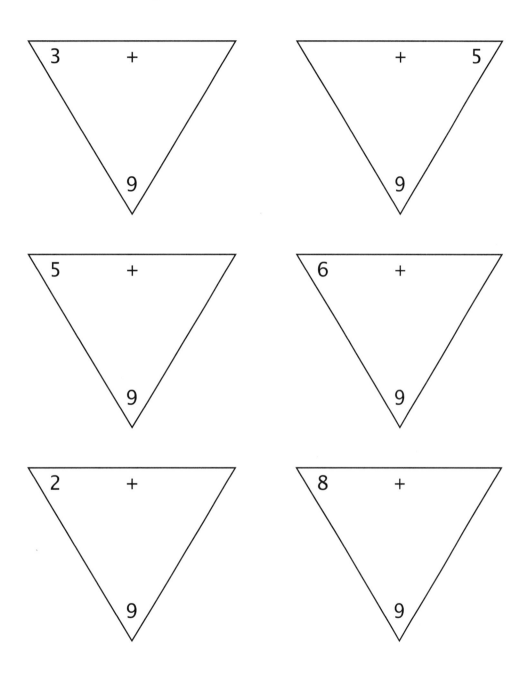

The triangles may be cut out and laminated if you wish. See the teaching tip in lesson 12 for more on using trios to teach addition.

EXTRA FUN 16X

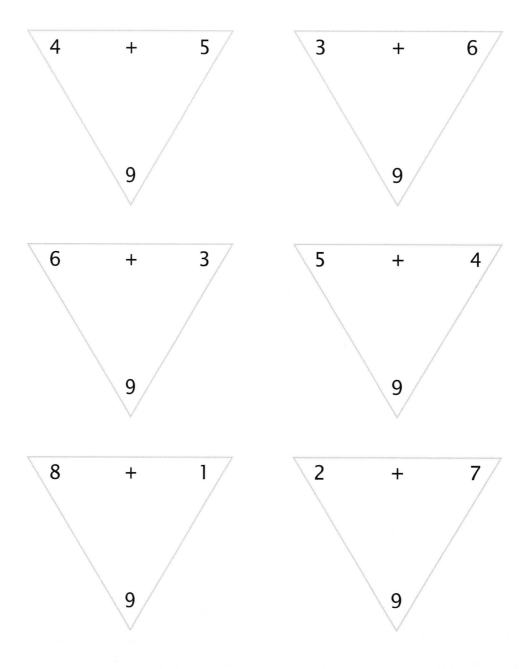

17X

Color the picture to practice the extras.

If the answer is 8, color the space yellow.
If the answer is 11, color the space purple.
If the answer is 12, color the space blue.

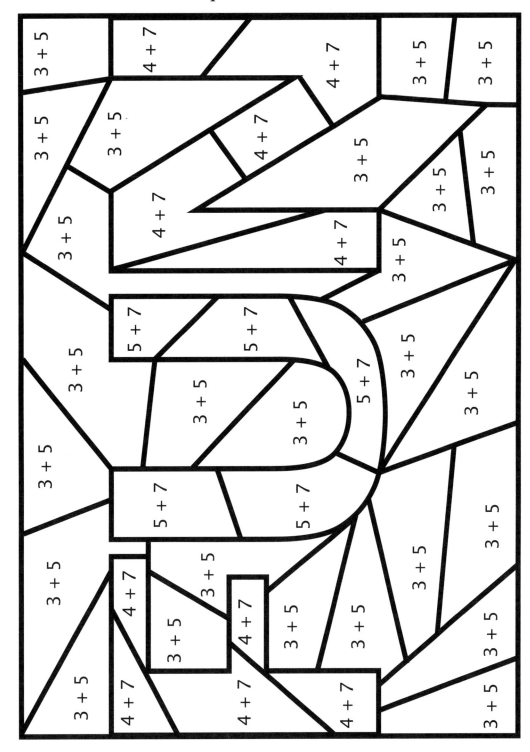

ALPHA EXTRA FUN

EXTRA FUN 17X

Fill in the blanks to write your own word problems for the extras.

1. _____ had three _____.

   A friend found five more of them.

   How many are there now? _____

2. _____ ate four _____

   and seven _____.

   How many things were eaten? _____

3. Five _____ came in the door.

   Seven of them jumped in the window.

   How many are in the room altogether? _____

Draw pictures to help you add and subtract.

2 cookies plus 3 cookies equals _____ cookies.
Draw the cookies.

$$\begin{array}{r} 2 \\ +\ 3 \\ \hline \end{array}$$

Sam ate 3 cookies.
Put Xs on 3 cookies in your picture.

$$\begin{array}{r} 5 \\ -\ 3 \\ \hline \end{array}$$

5 cookies minus 3 cookies equals _____ cookies.

---

4 cars plus 2 cars equals _____ cars.
Draw the cars.

$$\begin{array}{r} 4 \\ +\ 2 \\ \hline \end{array}$$

2 cars drove away.
Put Xs on 2 cars in your picture.

$$\begin{array}{r} 6 \\ -\ 2 \\ \hline \end{array}$$

6 cars minus 2 cars equals _____ cars.

EXTRA FUN 18X

Draw pictures to help you add and subtract.

5 candies plus 2 candies equals _____ candies.
Draw the candies.

$\begin{array}{r} 5 \\ +\,2 \\ \hline \end{array}$

Ethan ate 5 candies.
Put Xs on 5 candies in your picture.

$\begin{array}{r} 7 \\ -\,5 \\ \hline \end{array}$

7 candies minus 5 candies equals _____ candies.

4 trees plus 4 trees equals _____ trees.
Draw the trees.

$\begin{array}{r} 4 \\ +\,4 \\ \hline \end{array}$

4 trees were chopped down.
Put Xs on 4 trees in your picture.

$\begin{array}{r} 8 \\ -\,4 \\ \hline \end{array}$

8 trees minus 4 trees equals _____ trees.

19X

See if you can start at 20 and connect the dots by counting backwards by one. If this is too hard, start at one and connect the dots. Then start at 20 and trace the line backwards. Say the numbers as you go.

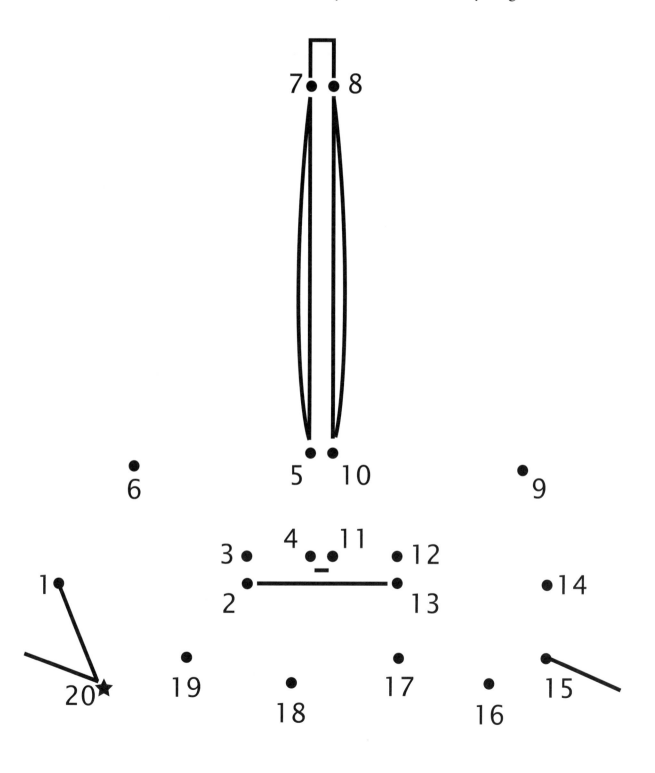

ALPHA EXTRA FUN

EXTRA FUN 19X

Do you like to visit your cousins? Did you know that the subtraction facts have cousins?

Fill in the boxes to show the cousin of each subtraction fact. The first one is done for you.

```
  5        5
- 4      - 1
---      ---
  1        4
```

```
  3        3
- 1      - 2
---      ---
  2        ☐
```

```
  6        6
- 0      - 6
---      ---
  6        ☐
```

```
  8        8
- 7      - 1
---      ---
  1        ☐
```

```
  4        4
- 1      - 3
---      ---
  3        ☐
```

```
  9        9
- 1      - 8
---      ---
  8        ☐
```

20X

See if you can start at 40 and connect the dots by counting backwards by two. If this is too hard, start at two and connect the dots. Then start at 40 and trace the line backwards. Say the numbers as you go.

Connect the dots from 40 to 2 counting backwards by two.

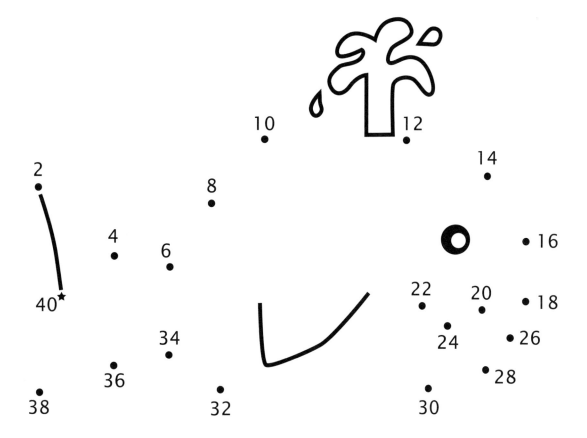

ALPHA   EXTRA FUN

EXTRA FUN 20X

Here are some more cousins who want to visit.

Fill in the boxes to show the cousin of each subtraction fact. The first one is done for you.

```
  5        5              7        7
- 2      - [3]          - 2      - [5]
---      ---            ---      ---
  3       [2]           [ ]      [ ]


 10       10              6        6
- 2      - [8]          - 2      - [ ]
---      ---            ---      ---
 [ ]      [ ]           [ ]      [ ]


  8        8             11       11
- 2      - [ ]          - 2      - [ ]
---      ---            ---      ---
 [ ]      [ ]           [ ]      [ ]
```

21X

Here is a challenge!

Start at 50 and count backwards by one.
Connect the dots from 50 to 1.

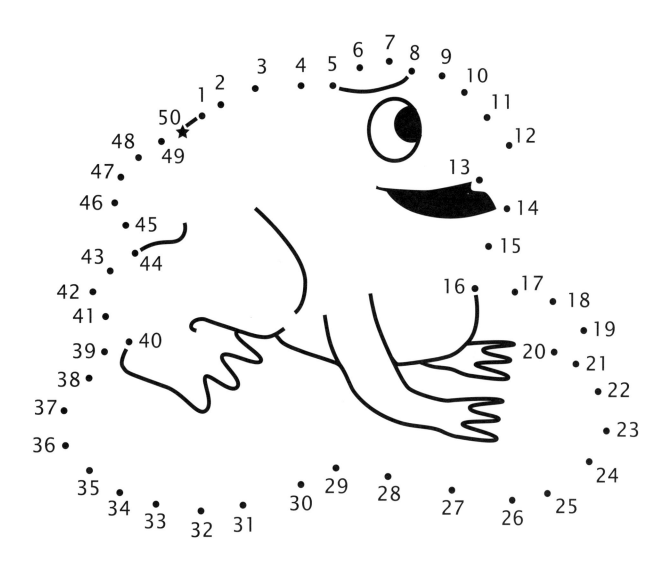

ALPHA EXTRA FUN 49

Word problems are fun! Use the blocks if you need to.

1. Sam did 9 math problems before lunch and 9 problems after lunch. How many problems did he do in all?

   _____

2. Chuck has 6 goldfish. Mike has 8 goldfish. How many more goldfish does Mike have?

   _____

3. Terri is 16 years old and Judy is 9 years old. What is the difference in their ages?

   _____

4. Ann colored 6 pictures in the morning and 7 pictures in the afternoon. She gave 9 pictures to her grandma. How many pictures are left?

   _____

5. Jeremy visited a farm. He saw 2 horses, 5 cows, and 8 sheep. How many animals did he see?

   _____

6. Mom bought 10 bananas. She used 2 bananas for muffins. How many bananas are left?

   _____

Color the picture to practice the subtraction facts you have learned so far.

If the answer is 3, color the space red.
If the answer is 4, color the space blue.
If the answer is 5, color the space green.
If the answer is 6, color the space brown.

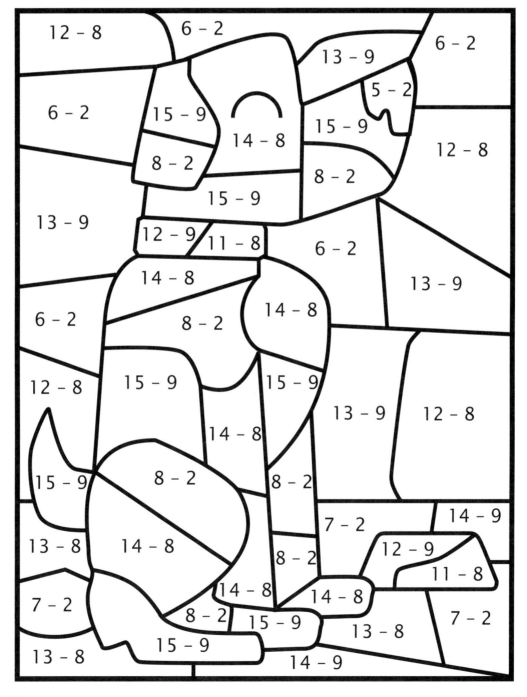

ALPHA EXTRA FUN

EXTRA FUN 22X

Fill in the boxes to show the cousins of these minus eight subtraction facts. The first one is done for you.

```
  11       11              14       14
-  8      -[3]            -  8     -[6]
----      ----            ----     ----
   3      [8]             [ ]      [ ]


  13       13              17       17
-  8      -[ ]            -  8     -[ ]
----      ----            ----     ----
  [ ]     [ ]             [ ]      [ ]


  15       15              12       12
-  8      -[ ]            -  8     -[ ]
----      ----            ----     ----
  [ ]     [ ]             [ ]      [ ]
```

23X

Subtract and use number words to write the answers in the boxes. The first one is done for you.

Across

1. 4 − 2 = __2__
2. 16 − 8 = ___
3. 14 − 7 = ___
5. 10 − 5 = ___

Down

1. 6 − 3 = ___
2. 2 − 1 = ___
3. 12 − 6 = ___
4. 18 − 9 = ___
5. 8 − 4 = ___

EXTRA FUN 23X

Here is a fun way to review your addition facts. Look at the example.

1. Add the numbers down each column and write your answers.

2. Add the numbers across each row and write your answers.

3. Add the numbers across the bottom and write the answer.

4. Now add the numbers in the right hand column. Your answer should match the answer already in the box!

#1
2	3	
2	3	
4	6	

#2
2	3	5
2	3	5
4	6	

#3
2	3	5
2	3	5
4	6	10

#4
2	3	5
2	3	5
4	6	10

Here are two new ones for you to try.

4	5	
4	3	

3	5	
4	1	

Practice the subtraction facts by finding the numbers that make ten. The triangles may be cut out and laminated if you wish. The answers are on the back of each triangle.

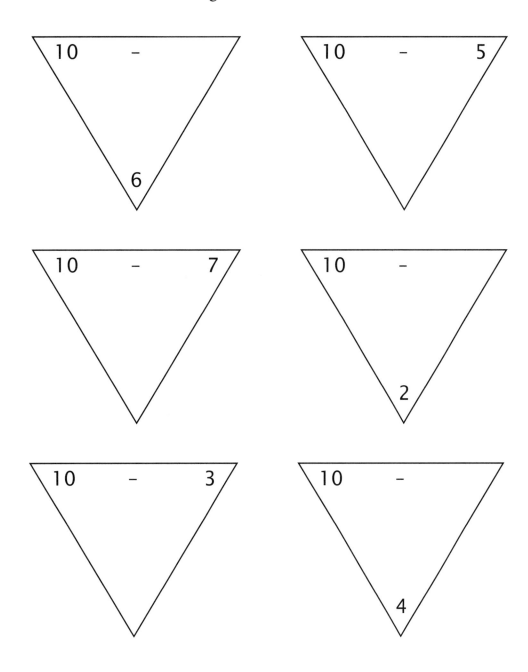

ALPHA EXTRA FUN

EXTRA FUN 24X

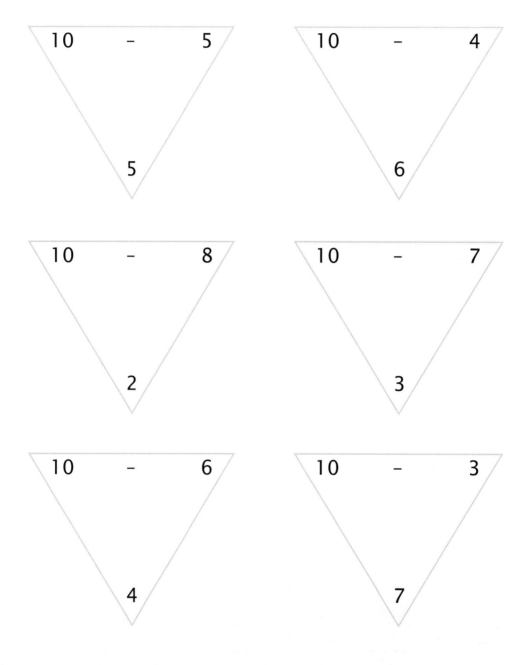

25X

Practice the subtraction facts by finding the numbers that make nine. The triangles may be cut out and laminated if you wish. The answers are on the back of each triangle.

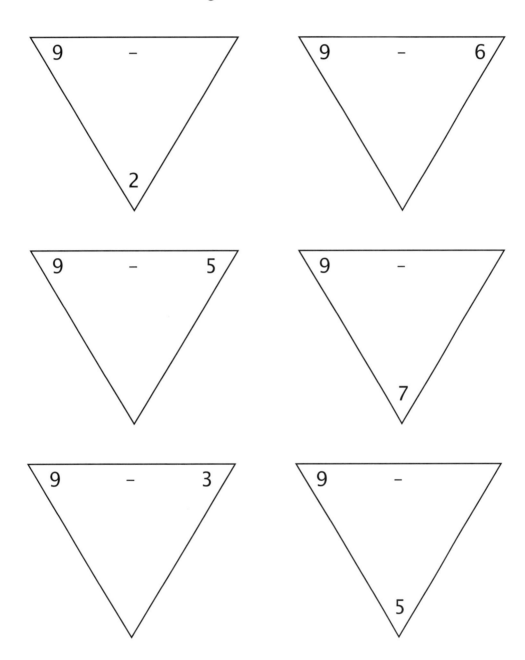

EXTRA FUN 25X

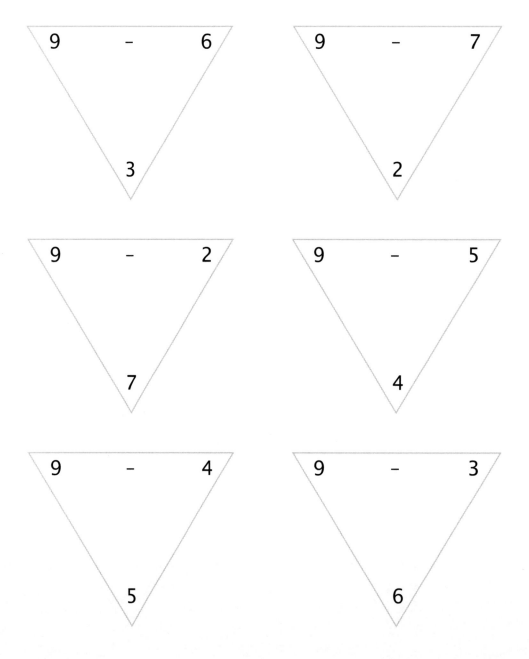

26X

Reviewing addition facts will help you learn your subtraction facts. Add across and down.

4	3	
5	2	

5	1	
2	7	

2	1	
3	4	

4	5	
4	1	

ALPHA EXTRA FUN

EXTRA FUN 26X

If the answer is 3, color the space blue.
If the answer is 4, color the space black.
If the answer is 5, color the space green.

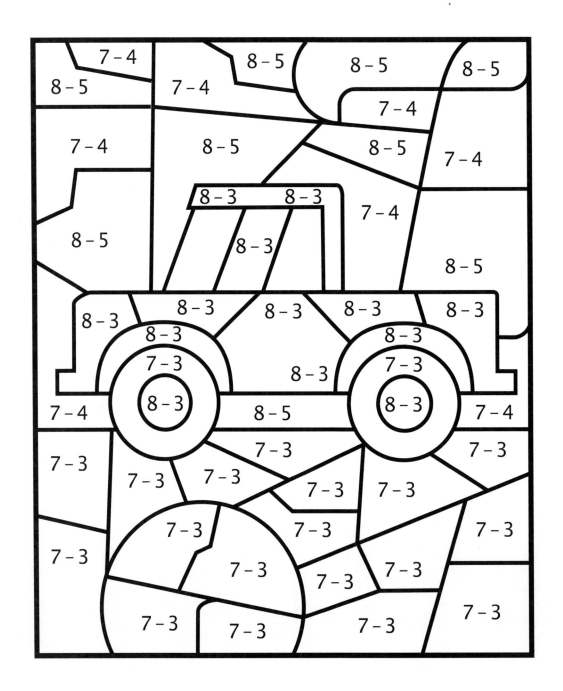

Skip count by 2 to 50 and connect the dots.

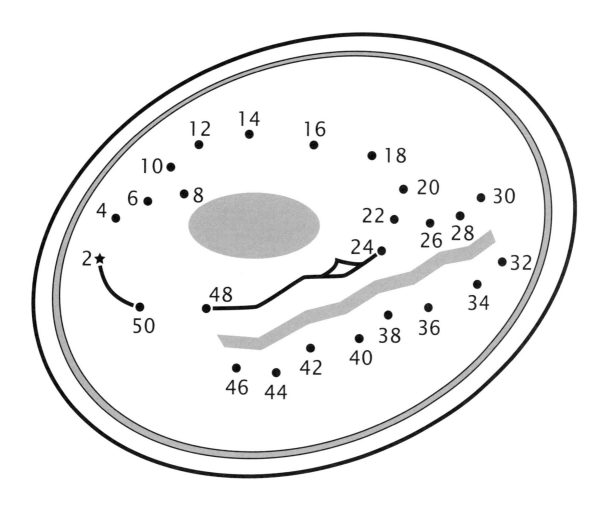

Word problems are fun! Use the blocks if you need to.

1. Karen and her friends ordered 13 ice cream sundaes. She and her friends have eaten 7 of them so far. How many sundaes are left?

   _____

2. Dennis did one math page a day for 5 days. After resting, he did a page every day for 6 more days. How many pages did he do in all?

   _____

3. Susan needs 11 birthday cards. She made 3 cards yesterday and 6 cards today. How many more cards does she need?

   _____   _____

4. Jared had 15 pennies. He lost 7 of them climbing a tree and 5 of them crawling under the porch. How many pennies does Jared have left?

   _____   _____

5. Kim walked 2 miles, ran 1 mile, and rode her bike 4 miles. How far did Kim travel?

   _____   _____

28X

Reviewing addition facts will help you learn your subtraction facts. Add across and down.

4	5	
1	1	

3	3	
4	2	

5	2	
4	4	

7	1	
2	4	

EXTRA FUN 28X

Use a yellow crayon to color all the boxes that have numbers that you say when you count by fives. See if you can go all the way to 100.

Next, use your blue crayon to color all the boxes that have numbers that you say when you count by tens. What color are the tens boxes now?

0	1	2	3	4	5	6	7	8	9
10	11	12	13	14	15	16	17	18	19
20	21	22	23	24	25	26	27	28	29
30	31	32	33	34	35	36	37	38	39
40	41	42	43	44	45	46	47	48	49
50	51	52	53	54	55	56	57	58	59
60	61	62	63	64	65	66	67	68	69
70	71	72	73	74	75	76	77	78	79
80	81	82	83	84	85	86	87	88	89
90	91	92	93	94	95	96	97	98	99
100									

If the answer is 5, color the space brown.
If the answer is 7, color the space green.
If the answer is 8, color the space yellow.
If the answer is 9, color the space red.

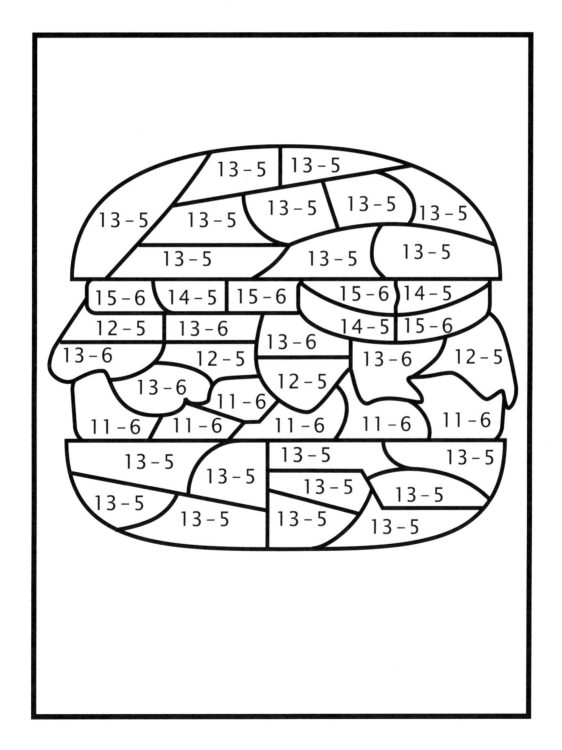

ALPHA EXTRA FUN 65

EXTRA FUN 29X

Color all the boxes that have numbers that you say when you count by twos. Go all the way to 100. You may use just one color or choose different colors to make a pattern.

0	1	2	3	4	5	6	7	8	9
10	11	12	13	14	15	16	17	18	19
20	21	22	23	24	25	26	27	28	29
30	31	32	33	34	35	36	37	38	39
40	41	42	43	44	45	46	47	48	49
50	51	52	53	54	55	56	57	58	59
60	61	62	63	64	65	66	67	68	69
70	71	72	73	74	75	76	77	78	79
80	81	82	83	84	85	86	87	88	89
90	91	92	93	94	95	96	97	98	99
100									

Start at zero and draw a line to the two. Keep counting by twos and drawing the lines to make a design.

Next, start at zero again and count by fives. Draw the lines to make a different design.

What happens if you start at zero and count by tens on the circle?

You may color your designs if you wish.

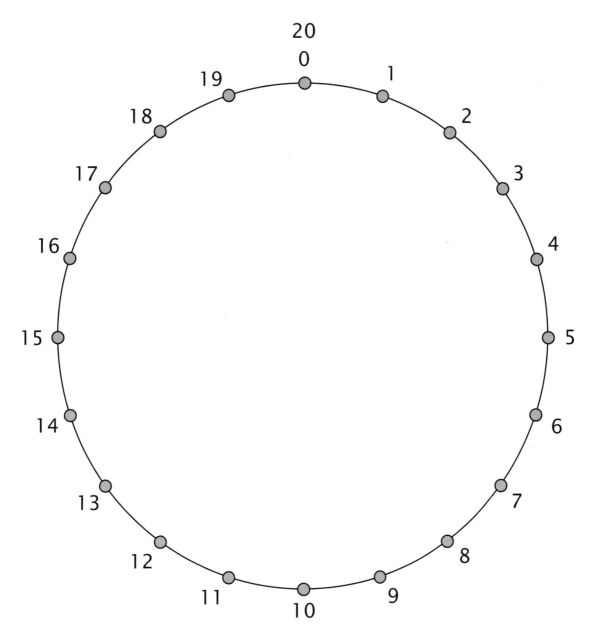

EXTRA FUN 30X

You should be a word problem expert by now!

1. Cara's dog had 11 puppies. She gave six of the puppies away. How many puppies are left?

   _____

2. Brandon had eight cookies. He lost three cookies and his dog ate one cookie. How many cookies are left?

   _____

   _____

3. Jessie has two red hats, five green hats, and six purple hats. How many hats does she have in all?

   _____

   _____

4. Amy has saved five dollars for a new game. The game costs 14 dollars. How many more dollars does Amy need to save?

   _____

5. Joanna bought 12 roses. Three of them were red and the rest were yellow. How many roses were yellow?

   _____

6. Four boys and seven girls wanted to get on the ride at the fair. There was only room for five children on the ride. How many children had to wait?

   _____

   _____

LESSON TESTS

TEST

1

Turn the paper sideways. Color the right number of blocks and say the answer.

1.

TEST 1

Count and write the number, and then say it.

2.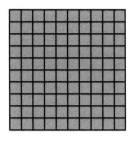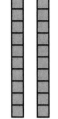

_____   _____   _____

Build and say the number.

3. 141

4. 390

5. What is the biggest number of units that can live in the units house?

_____

6. What is the biggest number of tens that can live in the tens house?

_____

TEST

2

Count and write from zero to twenty. Some numbers are already written for you.

1.

___ ___ ___ ___ ___ ___ ___ ___ ___ ___

_10_ ___ ___ ___ ___ ___ ___ ___ ___ ___

___

2.

___ ___ ___ ___ ___ ___ ___ _7_ ___ ___

___ ___ ___ ___ ___ ___ ___ ___ ___ ___

___

ALPHA TEST 2

TEST 2

Count and write the number, and then say it.

3.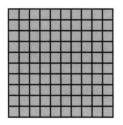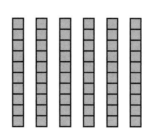

   _____   _____   _____

Build and say the number.

4. 157

5. 38

TEST 3

Match and say. Color the squares to match the blocks.

1.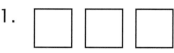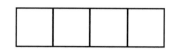

2.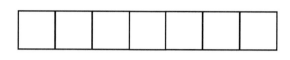

3.

4.

ALPHA TEST 3

TEST 3

Write or say the answer.

5. What color is the eight block? _____

Count and write the number, and then say it.

6.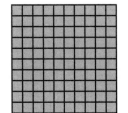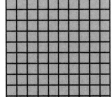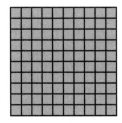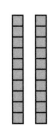

   _____   _____ _____

Count and write from zero to twenty. Some numbers are already written for you.

7.

___ 1 ___ ___ ___ ___ ___ ___ ___ ___

___ ___ ___ ___ ___ ___ ___ 18 ___

___

TEST 4

Solve.

1.   0
    + 1
   ———

2.   8
    + 0
   ———

3.   0
    + 6
   ———

4.   5
    + 0
   ———

5.   4
    + 0
   ———

6.   0
    + 3
   ———

7. 0 + 0 = _____

8. 0 + 7 = _____

9. 2 + 0 = _____

10. 1 + 0 = _____

11. 3 + 0 = _____

12. 0 + 5 = _____

TEST 4

Build and say the number.

13. 350

14. 102

15. What color is the nine block? _____

16. David ate 6 pieces of candy. His dad said, "No more!" How much candy did David eat?

_____

TEST

Solve.

1. $\begin{array}{r} 5 \\ +\ 1 \\ \hline \end{array}$

2. $\begin{array}{r} 4 \\ +\ 1 \\ \hline \end{array}$

3. $\begin{array}{r} 1 \\ +\ 8 \\ \hline \end{array}$

4. $\begin{array}{r} 2 \\ +\ 1 \\ \hline \end{array}$

5. $\begin{array}{r} 7 \\ +\ 1 \\ \hline \end{array}$

6. $\begin{array}{r} 1 \\ +\ 3 \\ \hline \end{array}$

7. 1 + 1 = _____

8. 1 + 9 = _____

9. 6 + 1 = _____

10. 0 + 5 = _____

11. 8 + 0 = _____

12. 0 + 2 = _____

TEST 5

Count and write the number, and then say it.

13.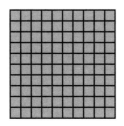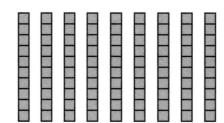

   _____   _____   _____

Build and say the number.

14. 214

15. There are five girls and one boy in Jen's family. How many children are there in all?

   _____

TEST 6

Count and write from 0 to 100.

1.

   0 __ __ __ __ __ __ __ __ __

__ __ __ __ __ __ __ __ __ 19

__ __ 22 __ __ __ __ __ __ __

__ __ __ __ __ 35 __ __ __ __

40 __ __ __ __ __ __ __ __ __

__ __ __ __ __ __ __ __ __ __

__ __ __ __ __ __ __ 67 __ __

__ 71 __ __ __ __ __ __ __ __

__ __ __ __ __ __ __ __ __ __

__ __ __ 93 __ __ __ __ __ __

__

ALPHA TEST 6

TEST 6

Skip count by 10 and write the numbers.

2. ___, ___, 30, ___, ___, ___, 70, ___, ___, ___

Solve.

3.   1
  + 6
_____

4.   5
  + 1
_____

5.   9
  + 0
_____

6.   1
  + 8
_____

TEST

Solve.

1.
$$\begin{array}{r} 1 \\ +2 \\ \hline \end{array}$$

2.
$$\begin{array}{r} 2 \\ +4 \\ \hline \end{array}$$

3.
$$\begin{array}{r} 20 \\ +20 \\ \hline \end{array}$$

4.
$$\begin{array}{r} 100 \\ +100 \\ \hline \end{array}$$

5.
$$\begin{array}{r} 6 \\ +2 \\ \hline \end{array}$$

6.
$$\begin{array}{r} 5 \\ +2 \\ \hline \end{array}$$

7.
$$\begin{array}{r} 2 \\ +7 \\ \hline \end{array}$$

8.
$$\begin{array}{r} 4 \\ +2 \\ \hline \end{array}$$

TEST 7

9. 2 + 3 = _____

10. 2 + 6 = _____

11. 0 + 2 = _____

12. 1 + 7 = _____

13. 3 + 0 = _____

14. 6 + 1 = _____

15. 9 + 1 = _____

16. 0 + 4 = _____

17. 1 + 2 = _____

18. Mia bought 3 red pencils and 2 green pencils. How many pencils did she buy?

_____

TEST 8

Solve for the unknown. Use the blocks if needed.

1. _____ + 1 = 1   2. _____ + 0 = 4

3. _____ + 2 = 2   4. _____ + 0 = 8

5. _____ + 2 = 5   6. _____ + 1 = 7

7. _____ + 2 = 3   8. _____ + 1 = 6

9. _____ + 0 = 0   10. _____ + 1 = 9

11. _____ + 0 = 1   12. _____ + 2 = 4

TEST 8

Solve.

13. 6
    + 2
    ___

14. 7
    + 1
    ___

15. 30
    + 20
    ____

16. 9
    + 0
    ___

17. Jordan has 3 model cars. He wants to have 5 cars. How many more cars does he want?

    _____

18. There are 9 players on a baseball team. We have 8 players. How many more players do we need to make one team?

    _____

TEST

Solve.

1. $\begin{array}{r} 0 \\ +9 \\ \hline \end{array}$

2. $\begin{array}{r} 9 \\ +7 \\ \hline \end{array}$

3. $\begin{array}{r} 6 \\ +9 \\ \hline \end{array}$

4. $\begin{array}{r} 9 \\ +9 \\ \hline \end{array}$

5. $\begin{array}{r} 9 \\ +2 \\ \hline \end{array}$

6. $\begin{array}{r} 3 \\ +9 \\ \hline \end{array}$

7. 9 + 1 = _____

8. 8 + 9 = _____

9. 9 + 7 = _____

10. 4 + 9 = _____

11. 6 + 1 = _____

12. 7 + 2 = _____

TEST 9

Solve for the unknown. Use the blocks if needed.

13. _____ + 9 = 11

14. _____ + 2 = 6

15. _____ + 1 = 4

Skip count by 10 and write the numbers.

16. 10, ___ , ___ , ___ , 50, ___ , ___ , ___ , ___ , ___

17. Jed read 9 books last week and 8 books this week. How many books did he read in all?

  _____

18. Alexis has six dollars. She needs eight dollars to buy a game she wants. How many more dollars does she need?

  _____

TEST

Solve.

1.   8
    + 8
   ___

2.   8
    + 5
   ___

3.   0
    + 8
   ___

4.   2
    + 8
   ___

5.   8
    + 6
   ___

6.   8
    + 3
   ___

7.   80
    + 10
   ___

8.   7
    + 8
   ___

TEST 10

9. 1 + 7 = _____

10. 8 + 9 = _____

11. 2 + 5 = _____

12. 9 + 7 = _____

13. 1 + 3 = _____

14. 9 + 6 = _____

15. 4 + 9 = _____

16. 7 + 2 = _____

Solve for the unknown. Use the blocks if needed.

17. $X + 7 = 15$

18. $X + 9 = 18$

19. Kayla rode her bike 8 miles and walked 5 miles. How far did she travel?

    _____

20. Emily needs 17 beads to make a necklace. She has 9 beads. How many more does she need?

    _____

UNIT TEST

Solve.

1. 7
    $\underline{+\ 9}$

2. 2
    $\underline{+\ 2}$

3. 4
    $\underline{+\ 9}$

4. 2
    $\underline{+\ 5}$

5. 6
    $\underline{+\ 2}$

6. 8
    $\underline{+\ 7}$

7. 20
    $\underline{+\ 10}$

8. 8
    $\underline{+\ 0}$

UNIT TEST I

9.   2
   + 9
   ———

10.   9
    + 9
    ———

11. 70
   + 20
   ———

12. 100
   + 800
   ———

13. 5 + 8 = _____

14. 8 + 4 = _____

15. 8 + 8 = _____

16. 2 + 4 = _____

17. 8 + 6 = _____

18. 6 + 9 = _____

19. 2 + 3 = _____

20. 9 + 5 = _____

21. 2 + 8 = _____

22. 9 + 9 = _____

23. 3 + 9 = _____

24. 9 + 8 = _____

UNIT TEST I

Solve for the unknown. Use the blocks if needed.

25. X + 7 = 15   26. X + 9 = 18

27. X + 8 = 17

Count and write the number, and then say it.

28.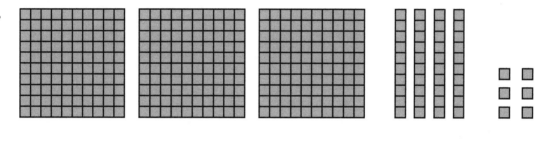

_____   _____ _____

Count and write from zero to twenty. Some numbers are already written for you.

29.

<u> 0 </u> __ __ __ __ __ __ __ __ __

<u> 20 </u>

TEST I

Skip count by 10 and write the numbers.

30. 10, ___, ___, ___, 50, ___, ___, ___, ___, ___

31. I saw 8 robins and 3 blue jays in my yard. How many birds did I see?

   _____

32. Michael found five pennies under the bed and one penny in the closet. How many pennies does Michael have?

   _____

TEST 11

Use the drawing to answer the questions.

1. How many striped circles are there? _____

2. How many plain triangles are there? _____

3. How many plain circles are there? _____

4. How many striped triangles are there? _____

TEST 11

Solve.

5. 9
   + 8
   ___

6. 8
   + 5
   ___

7. 2
   + 7
   ___

8. 40
   + 20
   ____

9. 3 + 0 = ____

10. 4 + 2 = ____

11. 9 + 9 = ____

12. 6 + 2 = ____

Solve for the unknown. Use the blocks if needed.

13. A + 8 = 14

14. 2 + X = 7

Skip count by two and write the numbers.

15. 2, ___, ___, ___, 10, ___, ___, ___, ___, ___

TEST

Solve.

1. 8
   + 8
   ----

2. 5
   + 5
   ----

3. 30
   + 30
   -----

4. 4
   + 4
   ----

5. 6
   + 6
   ----

6. 9
   + 9
   ----

7. 7
   + 7
   ----

8. 2
   + 2
   ----

9. 5 + 8 = _____

10. 9 + 4 = _____

11. 3 + 2 = _____

12. 6 + 9 = _____

TEST 12

Solve for the unknown. Use the blocks if needed.

13. $B + 9 = 18$

14. $Y + 8 = 15$

15. $Q + 2 = 11$

Count the number of each shape in the box.

16. There are _____ circles.

17. There are _____ triangles.

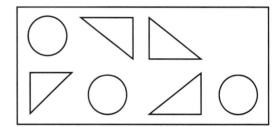

18. John went fishing with his dad. He caught 4 fish and his dad caught 4 fish. How many fish were caught in all?

    _____

19. Ali did 6 questions on her math worksheet. There are 15 questions in all. How many does Ali have left to do?

    _____

20. Chuck has 7 chores to do for his mom and 7 chores to do for his dad. How many chores does he have in all?

    _____

TEST

13

Use the drawing to answer the questions.

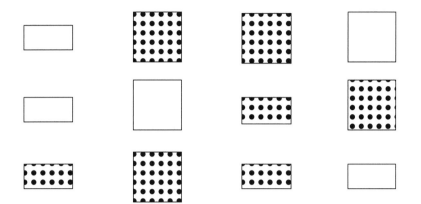

1. How many spotted rectangles are there? _____

2. How many plain squares are there? _____

3. How many plain rectangles are there? _____

4. How many spotted squares are there? _____

Solve.

5.   6
   + 6

6.   3
   + 3

ALPHA TEST 13

TEST 13

7.   7
   + 7
   ―――

8.   20
  + 40
  ―――

9. $9 + 0 =$ _____

10. $7 + 1 =$ _____

11. $8 + 9 =$ _____

12. $5 + 8 =$ _____

Solve for the unknown. Use the blocks if needed.

13. $A + 9 = 15$

14. $4 + X = 8$

15. Ashley drew a square on her paper.

   How many sides did it have? _____

Skip count by five and write the numbers.

16. ___, ___, ___, 20, ___, ___, ___, ___, 45, ___

TEST

Solve.

1. $\begin{array}{r} 50 \\ + 40 \\ \hline \end{array}$

2. $\begin{array}{r} 8 \\ + 7 \\ \hline \end{array}$

3. $\begin{array}{r} 1 \\ + 2 \\ \hline \end{array}$

4. $\begin{array}{r} 6 \\ + 7 \\ \hline \end{array}$

5. $\begin{array}{r} 2 \\ + 3 \\ \hline \end{array}$

6. $\begin{array}{r} 5 \\ + 6 \\ \hline \end{array}$

7. $\begin{array}{r} 8 \\ + 9 \\ \hline \end{array}$

8. $\begin{array}{r} 400 \\ + 300 \\ \hline \end{array}$

9. 7 + 7 = ____

10. 8 + 3 = ____

11. 6 + 6 = ____

12. 9 + 5 = ____

ALPHA TEST 14

TEST 14

Write the answers in words.

13. Three plus two equals _____ .

14. Two plus two equals _____ .

zero	six
one	seven
two	eight
three	nine
four	ten
five	

Circle the name of the shape.

15.
square   rectangle
circle   triangle

16.
square   rectangle
circle   triangle

17. Timothy has 5 toes on his left foot and 5 toes on his right foot. How many toes does Timothy have?

   _____

18. Caleb has five sisters and one brother.
    How many brothers and sisters does he have?

   _____

   Counting Caleb, how many children are in the family?

   _____

TEST

Solve.

1. $\begin{array}{r} 6 \\ +4 \\ \hline \end{array}$
2. $\begin{array}{r} 9 \\ +7 \\ \hline \end{array}$

3. $\begin{array}{r} 10 \\ +20 \\ \hline \end{array}$
4. $\begin{array}{r} 5 \\ +5 \\ \hline \end{array}$

5. $\begin{array}{r} 9 \\ +1 \\ \hline \end{array}$
6. $\begin{array}{r} 6 \\ +8 \\ \hline \end{array}$

7. $\begin{array}{r} 8 \\ +2 \\ \hline \end{array}$
8. $\begin{array}{r} 8 \\ +8 \\ \hline \end{array}$

TEST 15

9. 7 + 3 = _____

10. 3 + 4 = _____

11. 5 + 2 = _____

12. 6 + 4 = _____

13. 5 + 6 = _____

14. 7 + 8 = _____

15. 9 + 5 = _____

16. 7 + 6 = _____

Solve for the unknown. Use the blocks if needed.

17. $B + 6 = 10$

18. $Y + 5 = 10$

19. $Q + 3 = 10$

20. Lisa's mother said she had to pick up 10 toys in her room. Lisa has already picked up 8 toys. How many more toys does she have to pick up?

_____

TEST 16

Solve.

1.   2
 + 7
 ———

2.   5
 + 5
 ———

3.   10
 + 80
 ———

4.   6
 + 3
 ———

5.   7
 + 8
 ———

6.   5
 + 4
 ———

7.   6
 + 6
 ———

8.   5
 + 9
 ———

9. 8 + 1 = _____

10. 4 + 4 = _____

11. 8 + 4 = _____

TEST 16

Solve for the unknown. Use the blocks if needed.

12. 5 + A = 9

13. 6 + X = 9

14. 7 + G = 9

Match each shape with its name.

15.   triangle

16.   square

17.   rectangle

18.   circle

19. Sara can name 4 planets. If there are 9 planets in all, how many more does she have to learn?

    _____

20. Dan lost 60 pennies yesterday and 30 pennies today. How many pennies has he lost?

    _____

TEST

Solve.

1. $\begin{array}{r} 30 \\ +\ 50 \\ \hline \end{array}$

2. $\begin{array}{r} 4 \\ +\ 7 \\ \hline \end{array}$

3. $\begin{array}{r} 7 \\ +\ 5 \\ \hline \end{array}$

4. $\begin{array}{r} 5 \\ +\ 4 \\ \hline \end{array}$

5. $\begin{array}{r} 5 \\ +\ 7 \\ \hline \end{array}$

6. $\begin{array}{r} 7 \\ +\ 3 \\ \hline \end{array}$

7. $\begin{array}{r} 200 \\ +\ 200 \\ \hline \end{array}$

8. $\begin{array}{r} 8 \\ +\ 7 \\ \hline \end{array}$

TEST 17

9. 5 + 5 = _____   10. 9 + 9 = _____

11. 6 + 4 = _____   12. 3 + 8 = _____

13. 6 + 3 = _____   14. 8 + 5 = _____

Solve for the unknown. Use the blocks if needed.

15. 4 + X = 11   16. 7 + B = 12

17. 3 + R = 8

18. The guests brought six red balloons and five yellow balloons to Mary's party. How many balloons is that altogether?

_____

19. I noticed 4 bluebirds and 3 redbirds at my bird feeder. How many birds was that?

_____

Skip count by five and write the numbers.

20. 5, ____, ____, ____, ____, ____, ____, ____, ____, ____

UNIT TEST

Solve.

1. 9
   + 2
   ---

2. 3
   + 3
   ---

3. 7
   + 6
   ---

4. 3
   + 5
   ---

5. 5
   + 1
   ---

6. 2
   + 7
   ---

7. 40
   + 10
   ---

8. 2
   + 4
   ---

9. 2
   + 2
   ---

10. 2
   + 8

UNIT TEST II

Solve.

11. 50
    + 20
    ____

12. 400
    + 500
    _____

13.   8
    + 6
    ___

14.   9
    + 7
    ___

15. 40
    + 30
    ____

16. 300
    + 100
    _____

17. 8 + 3 = _____

18. 6 + 9 = _____

19. 6 + 6 = _____

20. 8 + 4 = _____

21. 0 + 6 = _____

22. 7 + 7 = _____

23. 5 + 6 = _____

24. 8 + 5 = _____

25.   7
    + 3
    ___

26.   9
    + 9
    ___

UNIT TEST II

Solve.

27. 9
   + 4
   ———

28. 7
   + 8
   ———

29. 4
   + 4
   ———

30. 2
   + 3
   ———

31. 20
   + 30
   ———

32. 7
   + 5
   ———

33. 0
   + 9
   ———

34. 5
   + 5
   ———

35. 10
   + 10
   ———

36. 600
   + 300
   ———

37.
```
   8
+  0
----
```

38.
```
   3
+  9
----
```

39.
```
   20
+  60
-----
```

40.
```
   300
+  300
------
```

41. 8 + 8 = _____

42. 4 + 7 = _____

43. 9 + 8 = _____

44. 4 + 6 = _____

Circle the name of the shape.

45.    square   rectangle   circle   triangle

46. square   rectangle   circle   triangle

47. 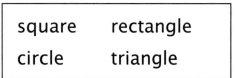   square   rectangle   circle   triangle

48. 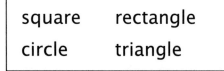   square   rectangle   circle   triangle

TEST 18

Identify each unit bar. Tell if it is have or owe. Color the bars if you wish.

1. ⬜⬜

   _____

2. ▣▣▣▣▣▣▣▣▣

   _____

3. ⬜⬜⬜

   _____

4. ⬜⬜⬜⬜

   _____

5. ⬜⬜⬜⬜⬜

   _____

6. ▣▣▣▣▣▣▣

   _____

ALPHA TEST 18

TEST 18

Solve.

7. $\;\;4$
    $\underline{+\;\;7}$

8. $\;\;6$
    $\underline{+\;\;5}$

9. $\;\;5$
    $\underline{+\;\;4}$

10. $\;600$
    $\underline{+\;300}$

Solve for the unknown. Use the blocks if needed.

11. X + 0 = 7

12. A + 1 = 3

13. Y + 0 = 6

14. B + 1 = 9

15. What number plus three is the same as eight?

    _____

16. What number plus six is the same as ten?

    _____

TEST

19

Subtract and write your answer. Check by adding up.

1. 4 − 1

2. 6 − 5

3. 8 − 8

4. 1 − 0

5. 2 − 1

6. 4 − 3

7. 7 − 0

8. 9 − 1

Solve.

9. 5 − 4 = ____

10. 9 − 8 = ____

11. 6 − 6 = ____

12. 10 − 1 = ____

TEST 19

13. 7 - 6 = _____

14. 6 - 1 = _____

15. 3 - 3 = _____

16. 0 - 0 = _____

Solve for the unknown. Use the blocks if needed.

17. Y + 7 = 7

18. R + 1 = 3

Choose the right answer and write it in the blank.

19. Ten minus one equals _____.

zero	six
one	seven
two	eight
three	nine
four	ten
five	

20. Chris found four dimes. He lost zero of them. How many dimes does Chris have now?

_____

TEST 20

Subtract find the difference. Check by adding up.

1.
$$\begin{array}{r} 9 \\ -2 \\ \hline \end{array}$$

2.
$$\begin{array}{r} 8 \\ -6 \\ \hline \end{array}$$

3.
$$\begin{array}{r} 3 \\ -2 \\ \hline \end{array}$$

4.
$$\begin{array}{r} 9 \\ -7 \\ \hline \end{array}$$

5.
$$\begin{array}{r} 5 \\ -2 \\ \hline \end{array}$$

6.
$$\begin{array}{r} 6 \\ -4 \\ \hline \end{array}$$

7.
$$\begin{array}{r} 7 \\ -2 \\ \hline \end{array}$$

8.
$$\begin{array}{r} 7 \\ -5 \\ \hline \end{array}$$

Solve.

9. 6 - 2 = _____

10. 10 - 8 = _____

11. 2 - 2 = _____

12. 7 - 1 = _____

TEST 20

Solve.

13.   7
    + 9
    ―――

14.   9
    + 2
    ―――

15.   4
    + 9
    ―――

16.   9
    + 9
    ―――

Solve for the unknown. Use the blocks if needed.

17. $B + 5 = 14$

18. $X + 8 = 17$

19. Eight children came to the picnic. Only two of them brought their baseball gloves. How many did not bring baseball gloves?

_____

20. Emily and her friend played a game. Emily made 7 points and her friend made 5 points. What was the difference between their scores?

_____

TEST

Solve.

1.   11
  − 9
  ―――

2.   10
  − 9
  ―――

3.   14
  − 9
  ―――

4.   17
  − 9
  ―――

5.   12
  − 9
  ―――

6.   13
  − 9
  ―――

7.   18
  − 9
  ―――

8.   16
  − 9
  ―――

9. 15 − 9 = _____

10. 9 − 9 = _____

11. 9 − 2 = _____

12. 8 − 6 = _____

TEST 21

Solve.

13. 5
   + 8
   ———

14. 8
   + 7
   ———

15. 4
   + 8
   ———

16. 800
   + 100
   ————

Solve for the unknown.

17. Q + 6 = 14

18. D + 8 = 11

19. Grace Joy counted 11 cars going by her house. Two of them were red. How many cars were not red?

   _____

20. David earned three dollars yesterday doing chores. Today he earned five dollars. How much more did he earn today than he earned yesterday?

   _____

TEST

Solve.

1. 12
    − 8
    ———

2. 15
    − 8
    ———

3. 13
    − 8
    ———

4. 14
    − 8
    ———

5. 16
    − 8
    ———

6. 10
    − 8
    ———

7. 17
    − 8
    ———

8. 9
    − 8
    ———

9. 15 - 8 = _____

10. 12 - 9 = _____

11. 16 - 9 = _____

12. 14 - 9 = _____

TEST 22

Solve.

13.   4
    + 4
    _____

14.   7
    + 7
    _____

15.   6
    + 6
    _____

16.   300
    + 300
    _____

Solve for the unknown.

17. $A + 2 = 9$

18. $X + 1 = 7$

19. Kim has a book with 11 chapters. She has read 8 chapters. How many are left to read?

    _____

20. Casey needs 17 dollars to buy a gift for her mom. She has 8 dollars. How many more dollars does she need?

    _____

TEST

Solve.

1. $\begin{array}{r} 10 \\ -5 \\ \hline \end{array}$

2. $\begin{array}{r} 2 \\ -1 \\ \hline \end{array}$

3. $\begin{array}{r} 14 \\ -7 \\ \hline \end{array}$

4. $\begin{array}{r} 60 \\ -30 \\ \hline \end{array}$

5. $\begin{array}{r} 4 \\ -2 \\ \hline \end{array}$

6. $\begin{array}{r} 16 \\ -8 \\ \hline \end{array}$

7. $\begin{array}{r} 12 \\ -6 \\ \hline \end{array}$

8. $\begin{array}{r} 8 \\ -4 \\ \hline \end{array}$

9. 18 − 9 = _____

10. 17 − 8 = _____

11. 12 − 8 = _____

12. 15 − 9 = _____

ALPHA TEST 23

TEST 23

Solve.

13.   4
    + 5

14.   9
    + 1

15.   3
    + 7

16.   8
    + 2

Solve for the unknown.

17. A + 5 = 10

18. B + 9 = 16

19. Hannah had 6 hours of school work to do. If she has finished 3 hours of work, how much is left to do?

_____

20. It is 12 miles to town. If I have 6 miles left to go, how many miles have I already traveled?

_____

TEST

24

Solve.

1.
$$\begin{array}{r} 10 \\ -3 \\ \hline \end{array}$$

2.
$$\begin{array}{r} 10 \\ -5 \\ \hline \end{array}$$

3.
$$\begin{array}{r} 10 \\ -2 \\ \hline \end{array}$$

4.
$$\begin{array}{r} 10 \\ -4 \\ \hline \end{array}$$

5.
$$\begin{array}{r} 10 \\ -6 \\ \hline \end{array}$$

6.
$$\begin{array}{r} 10 \\ -8 \\ \hline \end{array}$$

7.
$$\begin{array}{r} 10 \\ -7 \\ \hline \end{array}$$

8.
$$\begin{array}{r} 10 \\ -1 \\ \hline \end{array}$$

9. $16 - 8 = $ _____

10. $6 - 3 = $ _____

11. $14 - 7 = $ _____

12. $18 - 9 = $ _____

TEST 24

Solve.

13.   5
   + 4
   ―――

14.   7
   + 3
   ―――

15.   2
   + 7
   ―――

16.   3
   + 6
   ―――

Solve for the unknown.

17. $A + 1 = 9$

18. $B + 5 = 9$

19. Ethan is 10 years old and Luke is 4 years old. What is the difference in their ages?

_____

20. Joseph is planning to make 10 Christmas cards. If 7 are finished, how many more must he make?

_____

UNIT TEST

Solve.

1. 13 − 9

2. 11 − 8

3. 11 − 9

4. 14 − 9

5. 3 − 1

6. 8 − 4

7. 7 − 1

8. 4 − 0

9. 6 − 2

10. 9 − 9

UNIT TEST III

11. 9
   − 8
   ___

12. 10
   − 5
   ___

13. 3
   − 2
   ___

14. 7
   − 2
   ___

15. 15
   − 9
   ___

16. 2
   − 1
   ___

17. 16 − 9 = _____

18. 6 − 3 = _____

19. 10 − 9 = _____

20. 8 − 2 = _____

21. 18 − 9 = _____

22. 5 − 1 = _____

23. 17 − 8 = _____

24. 13 − 8 = _____

Solve.

25. 7
   − 5
   ———

26. 8
   − 7
   ———

27. 17
   − 9
   ———

28. 14
   − 8
   ———

29. 5
   − 3
   ———

30. 0
   − 0
   ———

31. 9
   − 2
   ———

32. 7
   − 6
   ———

33. 14
   − 7
   ———

34. 12
   − 9
   ———

TEST III

35.   6
    − 4
    ───

36.   12
    − 8
    ────

37.   6
    − 5
    ───

38.   16
    − 8
    ────

39.   5
    − 4
    ───

40.   2
    − 2
    ───

41. 6 - 0 = _____

42. 8 - 8 = _____

43. 8 - 6 = _____

44. 12 - 6 = _____

45. 4 - 3 = _____

46. 9 - 7 = _____

47. 13 - 8 = _____

48. 7 - 0 = _____

TEST

Solve.

1. 9
   $\underline{-4}$

2. 9
   $\underline{-7}$

3. 9
   $\underline{-6}$

4. 9
   $\underline{-0}$

5. 9
   $\underline{-2}$

6. 9
   $\underline{-3}$

7. 9
   $\underline{-1}$

8. 9
   $\underline{-8}$

9. $10 - 6 =$ _____

10. $16 - 8 =$ _____

11. $10 - 7 =$ _____

12. $14 - 7 =$ _____

TEST 25

Solve.

13. $\ 4$
    $+\ 3$

14. $\ 3$
    $+\ 5$

15. $\ 8$
    $+\ 5$

16. $\ 9$
    $+\ 4$

Solve for the unknown.

17. $Q + 1 = 5$

18. $R + 3 = 7$

19. There are 9 people on our team, but 4 of them are sick. How many people are left to play?

_____

20. Mom said that Mike could invite 9 people to his party. He has invited 6 people so far. How many more can he invite?

_____

TEST

26

Solve.

1. 7
   -3
   $\overline{}$

2. 8
   -5
   $\overline{}$

3. 7
   -4
   $\overline{}$

4. 8
   -3
   $\overline{}$

5. 9
   -5
   $\overline{}$

6. 9
   -6
   $\overline{}$

7. 15
   -8
   $\overline{}$

8. 70
   -30
   $\overline{}$

9. 12 − 9 = _____

10. 17 − 8 = _____

11. 14 − 7 = _____

12. 9 − 2 = _____

ALPHA TEST 26

TEST 26

Solve.

13. $\begin{array}{r} 7 \\ +\ 4 \\ \hline \end{array}$

14. $\begin{array}{r} 6 \\ +\ 7 \\ \hline \end{array}$

15. $\begin{array}{r} 7 \\ +\ 5 \\ \hline \end{array}$

16. $\begin{array}{r} 9 \\ +\ 7 \\ \hline \end{array}$

Solve for the unknown.

17. T + 7 = 15

18. P + 3 = 8

19. We got 7 inches of rain last month and 4 inches this month. How much more rain did we get last month than this month?

_____

20. Anna is 3 years younger than her brother. If her brother is 8, how old is Anna?

_____

TEST

27

Solve.

1. 16
   -7
   ___

2. 13
   -7
   ___

3. 15
   -7
   ___

4. 11
   -7
   ___

5. 12
   -7
   ___

6. 7
   -3
   ___

7. 8
   -5
   ___

8. 70
   -40
   ___

9. 8 - 3 = _____

10. 9 - 5 = _____

11. 10 - 6 = _____

12. 16 - 8 = _____

ALPHA TEST 27

TEST 27

Solve.

13.　　　5
　　　+ 6
　　　―――

14.　　　6
　　　+ 9
　　　―――

15.　　　8
　　　+ 6
　　　―――

16.　　　6
　　　+ 7
　　　―――

Solve for the unknown.

17.　Q + 4 = 11

18.　W + 9 = 18

19. Kay has 15 dollars in her pocket. If she spends 7 dollars, how much will she have left?

_____

20. Thirteen birds landed on the wire, and then seven flew away. How many birds are left?

_____

TEST 28

Solve.

1. $\begin{array}{r} 13 \\ -6 \\ \hline \end{array}$

2. $\begin{array}{r} 15 \\ -6 \\ \hline \end{array}$

3. $\begin{array}{r} 11 \\ -6 \\ \hline \end{array}$

4. $\begin{array}{r} 14 \\ -6 \\ \hline \end{array}$

5. $\begin{array}{r} 15 \\ -7 \\ \hline \end{array}$

6. $\begin{array}{r} 8 \\ -3 \\ \hline \end{array}$

7. $\begin{array}{r} 13 \\ -7 \\ \hline \end{array}$

8. $\begin{array}{r} 70 \\ -30 \\ \hline \end{array}$

9. 9 − 7 = _____

10. 8 − 1 = _____

11. 15 − 9 = _____

12. 11 − 8 = _____

ALPHA TEST 28

TEST 28

Solve.

13. 7
   + 5
   ___

14. 5
   + 9
   ___

15. 8
   + 5
   ___

16. 6
   + 5
   ___

Solve for the unknown.

17. A + 6 = 12

18. X + 4 = 13

19. Clara was washing the dishes. She washed 11 plates, but she broke 6 of them. How many plates are left?

   _____

20. Ben made 13 wooden toys to sell. If he has sold 6 of them, how many are left?

   _____

TEST

29

Solve.

1. 14
   -5
   $\overline{}$

2. 12
   -5
   $\overline{}$

3. 13
   -5
   $\overline{}$

4. 11
   -5
   $\overline{}$

5. 13
   -6
   $\overline{}$

6. 12
   -7
   $\overline{}$

7. 11
   -6
   $\overline{}$

8. 40
   -20
   $\overline{}$

9. 16 − 7 = _____

10. 14 − 6 = _____

11. 13 − 7 = _____

12. 15 − 6 = _____

ALPHA TEST 29

TEST 29

Solve.

13. 7
   + 4
   ———

14. 4
   + 9
   ———

15. 9
   + 3
   ———

16. 8
   + 3
   ———

Solve for the unknown.

17. D + 6 = 11

18. F + 8 = 12

19. Katie needed to learn 12 new words for science class. If she has learned 5 of them, how many are left to learn?

_____

20. The red picture book has 14 pages, and the green book has 5 pages. How many more pages does the red book have?

_____

TEST

Solve.

1. 12
   − 4
   ―――

2. 11
   − 3
   ―――

3. 13
   − 4
   ―――

4. 12
   − 3
   ―――

5. 11
   − 4
   ―――

6. 12
   − 5
   ―――

7. 11
   − 7
   ―――

8. 80
   − 50
   ―――

9. 11 − 5 = _____

10. 15 − 7 = _____

11. 14 − 6 = _____

12. 12 − 7 = _____

TEST 30

13. 13 - 5 = _____   14. 15 - 6 = _____

Solve.

15.   8   16.   6
     +   4   +   6
     _____   _____

17.   7   18.   9
     +   7   +   2
     _____   _____

19. Elizabeth did 13 math problems. She got 5 answers wrong. How many answers were right?

    _____

20. Maddie baked 11 cookies. Three of them got burned. How many cookies are not burned?

    _____

UNIT TEST

Solve.

1. 9
   -4
   $\overline{}$

2. 7
   -4
   $\overline{}$

3. 11
   -3
   $\overline{}$

4. 11
   -6
   $\overline{}$

5. 13
   -7
   $\overline{}$

6. 12
   -7
   $\overline{}$

7. 13
   -5
   $\overline{}$

8. 11
   -4
   $\overline{}$

UNIT TEST IV

9. 9
    -6

10. 7
    -3

11. 11
    -5

12. 13
    -6

13. 12
    -4

14. 8
    -3

15. 8
    -5

16. 11
    -7

UNIT TEST IV

17. 12
    − 3
    ─────

18. 15
    − 7
    ─────

19. 14
    − 5
    ─────

20.   9
    − 3
    ─────

21. 15 − 6 = _____

22. 16 − 7 = _____

23. 9 − 5 = _____

24. 14 − 6 = _____

25. 12 − 5 = _____

26. 13 − 4 = _____

UNIT TEST IV

FINAL TEST

Solve.

1.
```
   10
 -  3
 ----
```

2.
```
    7
 +  3
 ----
```

3.
```
    8
 -  4
 ----
```

4.
```
    4
 +  7
 ----
```

5.
```
    9
 -  6
 ----
```

6.
```
    9
 +  9
 ----
```

7.
```
   12
 -  7
 ----
```

8.
```
    8
 +  7
 ----
```

FINAL TEST

9. 15 	10. 12
   − 9 	 − 4

11. 5 	12. 13
   + 3 	 − 6

13. 10 	14. 7
   − 5 	 + 6

15. 3 	16. 11
   + 6 	 − 8

FINAL TEST

17. 8
   + 5
   ———

18. 4
   + 9
   ———

19. 17
   − 9
   ———

20. 14
   − 5
   ———

21. 3
   + 8
   ———

22. 13
   − 9
   ———

23. 5
   + 7
   ———

24. 16
   − 7
   ———

FINAL TEST

25.   9
    + 3

26.   11
    − 6

27.   15
    − 8

28.   7
    + 4

29.   5
    + 6

30.   8
    + 7